I0393518

RV Living

A Beginners Guide To Full-time RV Life & Achieving The Freedom Lifestyle

Jonathan Reid

Copyright © 2017 Jonathan Reid

All rights reserved.

ISBN-13: 978-1548466282
ISBN-10: 154846628X

© **Copyright 2017 – Jonathan Reid - All rights reserved.**

The contents of this book may not be reproduced, duplicated or transmitted without direct written permission from the author.

Under no circumstances will any legal responsibility or blame be held against the publisher for any reparation, damages, or monetary loss due to the information herein, either directly or indirectly.

Legal Notice: This book is copyright protected. This is only for personal use. You cannot amend, distribute, sell, use, quote or paraphrase any part or the content within this book without the consent of the author.

Disclaimer Notice: Please note the information contained within this document is for educational and entertainment purposes only. Every attempt has been made to provide accurate, up to date and reliable complete information. No warranties of any kind are expressed or implied. Readers acknowledge that the author is not engaging in the rendering of legal, financial, medical or professional advice.

The content of this book has been derived from various sources. Please consult a licensed professional before attempting any techniques outlined in this book.

By reading this document, the reader agrees that under no circumstances are is the author responsible for any losses, direct or indirect, which are incurred as a result of the use of information contained within this document, including, but not limited to, — errors, omissions, or inaccuracies.

JONATHAN REID

CONTENTS

DEDICATION

This book is dedicated to all my fellow dreamers, visionaries, achievers and people who refuse to remain and think inside the box. For those of us interested in redefining the term "wealth" from strictly just a measurement of money, to now include a person's life experiences and to be used as an overall measurement of ones state of happiness.

INTRODUCTION

Many people harbor secret dreams of dropping the 9-5 and hitting the road. This wild, adventurous lifestyle appeals to a deep, instinctual longing: to live meaningful, full lives, surrounded by beautiful natural environments, unburdened by society's expectations about how and where one should make their home.

Is it really such an unrealistic ambition? Despite millennials' collective concern about the instability of the economy, more and more of them seem to be taking the risk. Every day a new Instagram account or travel blog appears, encouraging modern dreamers to quit their day job and seek meaning and potential on the road. It seems that now, more than ever, it is trendy to be nomadic.

Many people worry about the practicalities of the lifestyle. It might sound like a lot of fun in theory, but how can you be sure that your favorite travel blogger is being completely honest in their portrayal of their daily life? After all, high-definition photographs of soaring mountains and majestic Redwoods don't tell the whole story.

If you're reading this book, you have taken the first step into making your dreams a reality. But it's important to be prepared. Which recreational vehicle (RV) should you buy? Where is it safe and legal to park your RV? How can you find community on the road? Where are the best spots to travel to? How can you make a living?

Even the most adventurous souls among us have questions, concerns, and doubts about the nomadic lifestyle. Whether you are contemplating becoming a road warrior sometime in the distant future or ready to take the leap tomorrow, this book compiles a few of the most common queries that newcomers to the RVing, camping, and traveling lifestyle have.

Of course, everyone's circumstances are different and your mileage may vary. (Yes, the pun was intentional!) This book is not meant to be a conclusive guide to RVing or living out of your RV. It is simply a brief introduction to the lifestyle that aims to dispel some persistent myths, offer insight, and encourage people.

I believe that this lifestyle is not only attainable but preferable for many people working in the gig economy, such as freelancers, temporary employees, stay-at-home moms, retirees, and recent grads who are feeling disillusioned about the current workforce. It is possible to see the world from the comfort of your vehicle and make money simultaneously. There has never been a better time to be a wanderer.

Hopefully, by the time you are finished this book, you will be excited about your future as a full-time RVer and knowledgeable enough that you can make your transition from the rat race to the road as seamless and stress-free as possible.

CHAPTER ONE:
PURCHASING YOUR
RECREATIONAL VEHICLE (RV)

When taking the leap and buying your first recreational vehicle, it makes sense that you might have a lot of questions. It is, understandably, an overwhelming prospect to many people, especially first-time buyers.

You may have heard horror stories from fellow travelers or read about deals gone wrong on your favorite camping blog. To ensure you don't end up living a similar nightmare, arm yourself with as much knowledge as possible before you hit the used-car lot or search on Craigslist for your vehicle.

As with any large purchase, you'll want to take precautions, be prepared, and know what you're getting into. A used RV can be inconvenient at best and deadly at worst; if there's something wrong with the vehicle, its current owners likely won't tell you—or at least, they won't be entirely honest.

What are some red flags to watch out? Anything that suggests the vehicle isn't in tip-top shape should ping your radar. For example, if the vehicle is missing parts, has noticeable water damage, or has been in storage for several years without being used, you might want to go elsewhere.

Repairs can be expensive, and not everyone is equipped to handle problems that may result from a damaged RV, especially if you're out on the road when it happens. Save yourself the trouble of inheriting someone else's problem unless you know enough about RV repairs to turn a fixer-upper into a flawless home on wheels.

Though we understand the appeal of finding a great bargain, it may be worth your while to spend a little extra on a vehicle in better shape than skimp and get stuck with an RV that breaks down on the freeway before you even past the state line. Save your savvy shopping habits for the future, when you're safely on the road.

What to Do Before You Buy

We advise that you review the following checklist before you go to look at a vehicle. Feel free to take a look at numerous vehicles before you make a decision. Your personal safety and security is at risk if you make a bad purchase. This is not a decision to be taken lightly!

Choose a model. Not every RV is created equal. Before you get intimidated by the many different models out there with weird-sounding names, keep in mind that there are a limited number of options available, and which one you choose depends mostly on one single question: driving or towing?

If you'll be driving, you need a motorhome. If you'll be towing your vehicle, you need a "towable" model. Easy as that! The most common types of recreational vehicles include:

- Travel Trailer
- Pop-Up
- Fifth Wheel
- Class A
- Class C

Once that decision has been made, you should find it fairly simple to narrow down your search.

Choose a floor plan. What confuses a lot of people about buying a motorhome is not the distinction between the models, but the distinction between the floor plans. There are just about as many options for floor plans as there are models, but you can't always see what a vehicle looks like from the outside.

There are three main types of interior floor plans, and which is best for you will depend on a few things: how many people live in the RV, how much entertaining you do, what large items you're carrying (such as motorcycles), and how much space you enjoy having to yourself.

No matter whether you prefer a spacious interior that reminds you of a luxury condo on wheels, or you like roughing it to the extreme, it is entirely possible to find a floor plan that meets your exact needs.

If you're travelling with kids or often bring friends aboard, a bunkhouse is the way to go; there are a set of bunk beds at the back. Toy hauler floor plans are designed specifically for travelers who are hauling around motorcycles and ATVs. If you are an extrovert, love to entertain, or simply like having the illusion of lots of space, a "rear living" floor plan is ideal. Many of them have an exterior cooking space, giving you the option of taking your party outside.

Do your research. Read customer reviews of the model you're buying, and don't be afraid to ask questions! If you are inquiring about a specific vehicle, find out who the official owner is. Sometimes, especially on unmonitored sites like Craigslist, friends and relatives will post ads and try to sell the vehicle for someone else. This means that they might not know anything about the state or condition of the vehicle and will not be able to give you a comprehensive, detailed review of the vehicle's history.

You are completely within your rights to ask for proof that your contact is the owner of the vehicle, and it is advisable to do so. It is also crucial that you ask to see maintenance records before you offer up a single dime. Sellers expect to answer lots of questions, so don't worry about bothering them. If they are deliberately vague or evasive about any of your questions, take heed. They might not be telling you the whole truth and the vehicle might not be in the best condition.

Use your connections. There's a good chance you or someone close to you knows someone who is selling a pre-owned vehicle, especially if you've been talking to other nomadic families in preparation for your upcoming transition to the RV lifestyle. Even 9-5ers often own RVs for their summer camping trips.

Maybe a friend of a friend has recently become an empty-nester and doesn't need to cart the kids to Yellowstone every summer anymore. If they're not making use of their vehicle, they might be happy to sell it for some extra cash. If you never ask, you'll never know.

Not to mention, if you buy your RV from a close friend or relative, you can pretty much guarantee they'll give you their honest opinion. They're not out to make a buck; they just want to get rid of the vehicle they're not making use of anymore. If you buy from a trusted acquaintance, they'll be able to answer any questions you have about the make, model, vehicle history, and previous trips.

If you don't know anyone who is currently selling, consider finding community among other RV "lifers." In today's world, many young, hip, and informed travellers share their experiences on social media and on their personal RVing blogs. Local communities exist, too, such as the Good Slam Club, and you can attend RV rallies in virtually every state. But in this day and age, it's great to take advantage of what the Internet has to offer.

Maybe you already follow a specific couple on Instagram, and you're in love with the model of RV they drive around the country, but you have no idea what it is or how to get one of your own. If you have questions about how to get started, message them privately or send them an email. Most RVers and campers are happy to share their insight and tell stories of what not to do at the beginning of your new life on the road. Finding mentors who you can trust and rely on is the key to getting off on the right track.

Ask the right questions. For instance, ask about the owner's last trip. If the vehicle has made it safely across the country three times in as many years, there's a good chance you'll be safe. Not to mention, a vehicle that has been driven frequently is very likely being well-maintained; lower mileage can actually be a red flag! Most importantly, ask how many previous owners have driven the vehicle, when the tires were last replaced, how many miles are on it, and if it has any minor issues.

While it is almost too easy to find used RVs online, not everyone is okay with the idea of buying from someone they don't know. If you are uncomfortable with the idea of purchasing from an individual, you may prefer to visit a licensed RV dealer in your area. They may not know as much about the vehicle's history, but they will almost certainly be aware of the last time the vehicle was repaired.

The NADA (National Automobile Dealers Association) will give you a good idea of every model's current going rate. Many people become intimidated because they're not sure what a fair price is for a specific model, or they jump too quickly because they're excited to buy.

Don't be that person! Lots of unlicensed dealers will try to take you for everything you're worth. Others will lie to you about the current market value of their vehicle. Take some time to research and determine what the most reasonable price is, and don't pay more than you can afford.

Inspect, inspect, inspect! Don't leave any stone unturned. Look at the exterior and interior of the vehicle thoroughly, and check for any visible water or weather damage. Check to ensure that every part of the vehicle is working properly—that includes the engine, electrical, mechanical, and plumbing. Some veteran campers suggest flushing the toilet and standing in the shower until it starts up before making a purchase.

The website **ChangingGears.com** offers a comprehensive checklist of things to review before you buy, and many people find it helpful to go down the list directly. You might forget to inspect one part of the vehicle when you go to see it, and end up paying for it later when it turns out that that important detail shouldn't have been overlooked. For example, it doesn't matter if the electricity is working if the plumbing isn't.

If there are small repairs needed, this isn't a huge problem, especially if you have some experience with car repairs to begin with. If you have very little experience making repairs on your own, take heed. Is this a risk you're willing to take? Talk it over with the owner and figure out a plan. Usually, if they know they're making a sale, a dealer will reasonably agree to fix repairs before handing the keys over to you. If they aren't able to fix the problem or direct you to someone who can, look elsewhere.

Keep up appearances. If you're going to be living in it long-term, you want to be satisfied with the appearance of your motorhome! Decide whether you want to keep the current interior aesthetic or renovate. It is advisable not to focus too much on this in the early stages, especially if you don't have a huge budget; you can always redecorate later on down the road.

First and foremost, you should care about the inner workings of the vehicle and the interior floor plan. Is it suitable for your family? If you are travelling by yourself, will you be safe? If a problem arises, do you know how to fix it or who to call? In the case of an emergency, the colour of the tiles over the sink isn't going to matter.

CHAPTER TWO:
BUDGETING, SAVING, AND
EARNING AN INCOME ON THE
ROAD

Once the big purchase has been made, you might be raring to go. We don't blame you. Both the North American and European countryside's are gorgeous, untainted landscapes, just waiting to be explored. But before you pull out the road atlas and set your sights on the next beachfront or mountainous locale of your dreams, there is one major factor you need to take into consideration before you head out anywhere: money.

One of the most frequently asked questions about RVing is, "How do you work while you're on the road?" Similarly, many people hold back on chasing their dream because of their fears of financial insecurity.

How are you going to earn an income? If you have money saved, how are you going to save it for the duration of your trip?

These questions are perfectly reasonable, and it makes sense that you're thinking practically. But remember, the whole point of this lifestyle is to leave the 9-5 behind—in the rearview, if you will. Your approach to personal finance doesn't have to fit the mould you've become accustomed to, and you don't have to define success by your yearly income.

On the contrary, success can be defined in a variety of ways. Only you get to decide what success means to you. Be strategic. Ask yourself a few more specific questions:

Can you work from the comfort of your vehicle? Do you have any artistic or crafty skills? Do you know how to "buy used and save the difference"? Are you thrifty, good at finding bargains? Are you thrilled by the prospect of being a freelance writer, or sharing your experiences on a personal travel blog?

You may have heard it said that you should never move to a brand-new city without at least three months' living expenses in your bank account. A similar rule applies to setting off in your airstream, motorhome, or camper van. If you don't have enough savings to keep yourself going for at least three months on the road, you know that you have to find a way to pull in funds to keep yourself going on the road.

Financial Benefits of RV Living

If you've sold your home and no longer have a mortgage, you'll be off to a great start. You'll have considerably more savings each month that you can put towards whatever your heart desires: recreational activities, hobbies, or even starting that small business you've always dreamed of.

If your credit card bills have been bumming you out, you

might find you're using them less since you actually have more cash up front. The rat race doesn't work for everyone. If you've chosen to make the transition to full-time RV living, you've probably already figured out that it's not working for you.

If you've never been a homeowner and have found monthly rent to be a crushing blow to your paycheck, living out of your camper van or motorhome will be a refreshing change. The only utilities you'll be paying are fuel—and you might have to pay for occasional repairs.

For many people, especially millennials, this alternative, off-the-grid lifestyle makes a lot of sense both financially and emotionally. Many people report feeling instantly happier and freer once they pay off their debts and make the transition to full-time RV living.

Managing Your Money

You don't have to stick rigidly to a budget every week or month. In fact, some people find setting a strict budget can be highly limiting and can lead to more excessive splurging. On the other hand, it is important to manage your money carefully, even if you're lucky enough to be financially secure when you set out in your motorhome.

Some tips for managing your money are below.

- Only use your credit cards for emergencies. Debt can collect quickly. We would recommend paying off *any* outstanding debts you have before you hit the road, including student loans, so you can truly start with a clean state... but if this is not possible, make strides to pay them off as quickly as possible before interest becomes astronomically high.

- Have a loose monthly budget and try to stick to it. If you go slightly over one month, don't worry. This happens to the best of us. However, it is wise to keep track of how much you're spending and what you're spending it on.

- Pay yourself first. Whenever you get paid, put a small percentage away into your savings account before doing anything else. This will ensure you aren't tempted to spend frivolously as soon as the money comes in.

- *Have* a savings account. Many Americans don't have any savings at all. If you fall into this category, make a point of starting to set aside a small amount of money each time you get paid. When you're living off the grid or travelling around in an RV, emergencies or unexpected expenses can easily crop up. You want to be prepared before it happens.

- Figure out what you can do without. Cut back on your daily $5 coffee and brew it in your motorhome each morning instead. We expect that after a few weeks on the road you'll find that you don't crave Starbucks as much anymore if the alternative is a freshly brewed coffee in the middle of a field full of spring flowers in upstate New York, a sprawling desert in Arizona, or a forest with a view of the majestic Rocky Mountains.

- Find simple pleasures in everyday life. The minimalist movement is intrinsically connected with the slow living movement, which appeals to many RV travellers. Even if you were a hoarder before you set out on the road, you'll quickly find that it's not possible to be one while living in a small space. Once you've started to appreciate the simple things, you won't desire more material items—and you'll save money.

- Shop secondhand. You may have heard the phrase "buy used and save the difference." It applies to pretty much everything. If you can buy it at a local Goodwill or find someone selling it cheaply on Craigslist, you shouldn't hesitate. Great bargains are everywhere, if you know where to look. Get in the habit of hitting up local thrift stores, garage sales, and antique shops. You never know what you might find.

- Stash change. If you have small change, keep it in a piggy bank or small jar. Label it with the name of the next place you'd like to visit, or a dream you'd like to achieve someday. Seeing the money gather over time, and reading the inspiring message on the front, will encourage you to keep working towards your goal.

Having a healthy attitude towards money helps encourage better financial decisions in the long term. If you're constantly stressing about it, you won't be able to enjoy the beautiful, natural views you're guaranteed to see on your trip.

It's important to be prepared for crises, especially if you're far from your home and family. If you put up safeguards, such as a savings account or a prepaid Visa card, before you ever get behind the wheel, you won't be worrying as much when an unexpected obstacle comes your way.

Freelancing and Paid Contract Work

If you're feeling discouraged about your personal finances right now, don't panic. In order to help you figure out how best to support yourself on the road, we've compiled a list of a few questions that will force you to figure out what your unique skill sets are—and if there is a way to make money using them online or in local camping communities.

Pull out a notebook, jot down these questions, and write the first thing that comes to mind.

Can you see yourself as a freelancer or an entrepreneur? The gig economy is thriving. It is estimated that by 2020, nearly half of all Americans will be temporary employees or contract workers. Many of them are beneath the age of 35, setting them firmly in the millennial category. Forbes 30 Under 30 details the amazing success stories of young entrepreneurs who said, "No thanks" to the traditional work lifestyle and made things happen.

While freelancing makes tax season a bit more complicated, and temps don't have the luxury of being unionized or receiving dental benefits, the freelance economy is experiencing a massive spike in popularity and growing at rates that are surprising economists.

It seems likely that this swift change in a new direction is an example of economic innovative in the face of a collapsing economy. Overworked and underpaid young people, frustrated by the difficulty they're facing finding high-paid work even with advanced degrees, are turning towards a market that appears to offer more potential for upward advancement. In the wake of the 2008 recession, young people are striving for success by moving against the current. Would you like to be a part of this inspiring new movement?

You'd be surprised to learn just how easy it is to make a decent income from the road. Plenty of people are living outside the confines of everyday society and thriving. Fortunately, in today's world, the Internet-savvy RVer is more than capable of making a living on the road. (Unless you've always been a really good saver, you'll be surprised at how much you're tempted to spend for the first few months on the road. After all, your mindset is still "I'm on vacation," not "I'm living this way permanently now.")

In fact, in some ways, this lifestyle offers *more* opportunities for financial growth than a steady, secure office job. A freelance job allows you a variety of unique, compelling opportunities that you would never get working a typical job:

- Flexibility. You can set your own hours! Work late at night, early in the morning, or whenever inspiration strikes. You can also set your own rates, and keep increasing them as you gain a larger portfolio.

- A break from the norm. You'll never have a dull moment as a freelancer. Each gig is different—some will be boring, some will be exciting. Each one will teach you something new.

- Network expansion. You'll be meeting new clients every day, making connections all over the country (or the world), and in the creative business, it's all about "who you know." For someone living out of their motorhome, the idea of having a friend to stay with in lots of far-flung, exotic locations can be immensely exciting.

- Free time. If you work hard, you'll have time to play even harder. Once you've met your deadlines for the day, you can take lots of time to do whatever makes you happy. Who knows? You might discover a new hobby or passion.

What professional skills do you already have? Even if you're a recent grad, you likely have a better CV than you realize. (And there are freelancers who will help you polish your resume, if you need help doing so!) If you've worked in retail, you have sales experience. If you've done your time in a clerical setting, you know how to use Word and

Excel and are probably a fast typist. If you've worked in a bank, you're good with numbers.

Maybe your psychology degree didn't get you your dream job straight out of college, or maybe you settled into a job you were overqualified for just to pay the bills. But chances are, that degree has offered you invaluable knowledge. A literature student could be a good writer. A sociology grad understands statistics.

If you've been in the workforce for many years, your resume likely includes much more information that you can use to your advantage. Every job you've ever had has led you to this point you're at now, and you can easily take something away from every experience.

What unique skills do you have? Now, think beyond your work and school experience. What did you used to love as a kid? What do you do in your spare time? Without the pressures of the traditional workforce to hold you back, you can pursue your dreams from the comfort of your vehicle.

Maybe you love taking photos and capturing memories on the road. Perhaps you wanted to study art, but started feeling as though a career in that field was impossible, so you changed your major.

Can you write well? Consider offering your services as a freelance writer, editor, proofreader, or ESL tutor. Do you speak more than one language? You could be a translator—a very highly sought-after role.

Are you a social media addict or blogger? You can monetize your blog by partnering with brands, doing sponsored posts on Instagram, or working as a freelance social media manager for companies.

Can you draw, sing, or play an instrument? Illustrators, session vocalists, and session musicians are in high demand. Sites like Slice the Pie allow you to review short music samples and get paid depending on the quality and length of your review.

Artisan Marketplaces and Online Stores

There's nothing people love more than discovering a new favourite brand, and small, boutique companies are becoming increasingly popular. If you've ever had the desire to make your own jewelry, design your own clothing, or start an online store, now is the time.

There is no better way to ensure you make an income from your handmade crafts, paintings, clothing, or accessories than by doing it from the road. Not only are you able to set up a store on an online marketplace like Redbubble or Etsy, you can also hit up every local festival while you're travelling.

If you already run an online store that is doing well, you could always do better. Think bigger: consider applying to be a vendor at a festival or convention in the next town you'll be visiting. If you make a one-on-one connection with people all over the country, word of mouth will spread and you'll start to draw in new shoppers.

Even in the Internet age, a good social media presence isn't all it takes to run a small artisan business. People appreciate the value of human connection. They like to talk to the designer or the artist face to face, get to know the story behind your business, and touch and hold the items. This is why it's important to find a community on the road.

Craft shows and artisan exhibitions are everywhere; you shouldn't have a problem finding a place to set up your booth and sell your wares. Best of all, you'll get to meet lots of new people that will make you feel welcomed. Even if you're an introvert, selling your stuff can be a great way to break you out of your comfort zone and force you to meet new and interesting people.

If you're still not convinced, open your laptop. Get inspired. Take a look at a few of your favourite travel blogs or online marketplaces. Read the book *The Four-Hour Workweek*. Google tips on how to get started making money online. Set up a PayPal or Venmo account. There are countless stories of people just like you who made the shift and are now making an income while living on the road. If it happened to them, why can't it happen to you?

CHAPTER THREE:
PERSONAL SAFETY AND HEALTH GUIDELINES

Once you've inspected your RV and are sure that it's in tip-top shape, you should keep in mind that dilemmas can still occur on the road. It's in your best interest to be prepared for some of the most common possibilities. If you're miles away from civilization and you get a flat tire, you could be in trouble.

On the other hand, parking in restricted or illegal zones overnight can get you a pretty steep fine that you will want to avoid. Take precautions before you hit the road so you can ensure you have a smooth ride all the way.

Accidents

According to Geico, the most common causes of accidents involving RVs are the following:

- **Tire blowouts.** Imagine you're driving through rough, uncertain terrain when suddenly a tire pops. Your vehicle spins out of control. What then? Ask yourself: when was the last time you checked your tires? Make sure they're new, properly inflated, and weather-resistant. If your dealer didn't provide you with updated tires when you purchased your vehicle, you're going to have to do this step yourself, but it could save you a ton of trouble down the line. Check the condition of your tires regularly and maintain the recommended pressure level to avoid unexpected tire blowouts.

- **Fires.** These are typically caused by leaks in the gas sources. Make sure you have your propane system carefully inspected before leaving on a trip, and check the exhaust system regularly. Does your RV have a generator? Leaving it running while you sleep puts you at high risk of carbon monoxide poisoning. Last but not least, install a fire alarm on board before you leave. Gas leaks can happen very quietly, so having an alarm warning you that something's amiss could mean the difference between life and death.

- **Overloading.** If you're a heavyweight traveler, you might want to consider lightening the load a bit. Uneven weight is the number one cause of tire stress and can also make steering and braking more difficult—something that can easily turn from inconvenient to deadly in an instant, if you're trying to swerve to avoid a collision. Water, fuel, and propane tanks increase the weight of your vehicle, so any additional cargo that weighs you down should be reconsidered. Don't overpack and take only what you need. Your RV's manufacturer should be able to tell you what the recommended maximum weight capacity is. Whatever you do, don't exceed it. If you're unsure, you can weigh your fully loaded vehicle using platform scales.

- **Height and clearance errors.** Do you have your RV's height and clearance information on hand? Stuff it in the glove compartment or next to your seat; you never know when it will be crucial. Most mistakes occur as a result of an RV driver not knowing their vehicle inside and out. If you know you will be driving under bridges and tunnels along your route, prepare beforehand and figure out if your RV can safely clear them. If they can't, figure out a different route. Some gas station overhangs are designed for smaller vehicles and hang especially low, so be wary of them as well.

- **Rodent infestations.** When you're living and cooking out of your RV, you're basically announcing to the local rodent population, "Come on in! Free food!" They may seem like only a minor inconvenience, but they can actually cause more harm than you might expect. Many species of rodents will go straight for loose wires and chew them until they snap in half. If you suspect you've got a little friend living aboard, regularly check your RV's wires to make sure they haven't been chewed straight through. Set humane traps or call an exterminator if you think you have a problem.

- **Insecurely stored cargo.** Make sure everything is safely secured long before you ever leave your driveway. This includes awnings, detachable steps, motorcycles, attachable trailers, and anything else you might be bringing along with you. And remember, if you think a piece of cargo is going to dangerously increase the overall weight of your vehicle, ask yourself, "Is it really necessary to bring this along?" You might be surprised what you're willing to part with when you realize that storing it is going to be a hassle.

Overnight Parking

One of the most commonly asked questions from new RVers is where they can safely park their vehicle overnight. Since people who live in their RVs are on the road constantly and always thinking ahead, the risk of forgoing sleep and powering through using caffeine and adrenaline can be tempting. Unfortunately, it can also be highly deadly. Statistically, most accidents occur when drivers haven't gotten enough sleep.

Since your vehicle is your home on wheels, it is especially risky to sacrifice sleep for adventure. If you don't know where your next camping spot will be, you might end up falling asleep at the wheel, or pulling over to sleep only to wake up to a hefty fine on your dashboard.

The good news? Finding a place to crash for the night that is safe, reliable, and legal isn't nearly as challenging as you might believe. Not many people know that a lot of large corporations actually permit overnight camping in their parking lots, and that highway rest stops are a surprisingly safe option.

Though the ideal option is always a traditional campground where you can mingle with likeminded people and be sure you'll be safe from potential break-ins, if you have to rough it in a non-traditional location, consider one of the following:

- Walmart. It is company policy to allow overnight campers in Walmart parking lots! Up to 80% of these superstores permit RV campers to stay overnight, and most veteran RVers will tell you they have had positive experiences camping in a Walmart lot. Be sure to alert the store manager to your presence before you take up residence, and they will pass the information along to store security. Not only do they welcome RV campers, they will actually keep an eye out for your safety.

- Camping World. The name might be misleading, but this is not a campground. It's actually one of the most popular chains of camping stores. Understandably, they are very RV-friendly, and open their parking lots to overnight campers at most locations across the United States. What's more, they offer RV repairs and maintenance onsite, so if anything goes wrong, you're only steps away from professional help.

- Flying J Truck Stops. Unlike other truck stops, this chain is particularly friendly to RV campers. They actually have designated areas in their parking lot for camper vans, and the food and gas is cheaper than at other truck stops. It is an unspoken rule not to park near the trucks. Be respectful of the long-haul truckers, who are often carrying heavy cargo and taking up a lot of space in their designated spots. Give them their space and they will give you yours.

- Kmart. Along with other, lesser known factory outlet stores and large-scale shopping centers, Kmart welcomes overnight camping, and most store managers will alert security that you're there. Kmart stores with Super Centers offer affordable meals, too.

- Elks Club Lodges. Most locations have RV hookups onsite, for either free or very cheap. If you're an Elks Club member, you can take advantage of all the amenities, including dining at the restaurants, drinking in the lounges, and joining other members for dancing and live music in the evenings. For travellers on the road, this can be a great way to meet new people and socialize. However, many Elks Club locations are starting to increase their camping fees due to the amount of people hoping to stay without a membership. Some will ask for a donation before permitting you to use their hookup.

Avoiding Crime

Sometimes, even the best laid plans go awry. If you talk to any veteran RVer for an extended period of time, you'll likely hear a few horror stories—about a bear lurking around a rural campground, a gunshot going off in an inner-city area, or a wild-eyed stranger approaching with a rifle to warn you to get off his property.

Chances are, you'll occasionally be forced to camp somewhere you don't feel one hundred percent comfortable. Even if you resolve to only camp in guarded Walmart parking lots and four-star campgrounds, you can never fully prepare for the unexpected. That's half the fun of roughing it, after all. But there are ways to avoid emergency situations and ensure you don't get caught off guard.

Firstly, do your research about a campground before you decide to stay there. For the most part, they're fenced in, patrolled by 24-hour security guards, and equipped with CCTV cameras, so it's unlikely that outsiders will even be able to find their way in. However, if you don't know the community, you can't be sure that your fellow campers are trustworthy, or that they will look out for your best interests.

If you're heading out to swim, hike, or get food, lock up your RV, no matter what. Get in the habit of doing this even if you'll only be out for a few minutes. Lock up the exterior and draw the curtains if you're especially concerned. (This also has the added benefit of preventing your fabrics from fading if your RV is parked in the sunlight.) Install devices that will alert you to any unusual activity. A motion-detector vehicle alarm system may occasionally go off in the middle of the night, but it's a small price to pay for yours and your family's safety.

If you have a trusted friend in the campground, ask them to keep watch. Truthfully, most RVers have an unspoken rule to look out for one another. We've all been there at one point or another, and as a result, we're all pretty street-smart and savvy travelers. If anyone comes prowling around looking for things to steal, they'll likely already have multiple sets of eyes on them.

Once you're more comfortable living the RV life, you might start hearing stories about "boondocking." More experienced adventurers will likely suggest that you try this at least once, as it provides a more authentic, earthy camping experience than staying at a resort-like campground where you're only a few miles down the round from civilization.

The tried-and-true RV lifers swear by boondocking, and it's understandable. In our fast-paced world, it can seem deeply appealing to have the opportunity to awaken beneath a wide open desert sky, cook on an open fire on a deserted beach, or stargaze without the steady interruption of noise and air pollution. But are you ready for it?

There are obvious risks involved in camping in the true, untouched wilderness. Camping in a forested area far off the beaten path is not a secure or reliable way to camp. Even if you're pulling off for a quick power nap in a forested enclave near a major highway, you will mark yourself out as a target since you're parked in an isolated area.

A good rule of thumb is to never open the door to anyone, unless they're a police officer; if someone is holding a badge, they're probably going to tell you you've chosen an illegal camping spot, or warn you about the possibility of crime. Trust the locals' opinions more than anything you may have read in a roadmap or heard through the grapevine. While some campers take impossible risks and almost seem to thrive on the possibility of dangerous, it's always better to be safe than sorry.

If you're planning to park your RV somewhere highly rural, isolated, or unknown, check local bylaws to make sure you're even allowed to. Just because the land doesn't appear occupied doesn't mean it doesn't have an owner. (In some European countries, the law states that anyone is allowed to camp anywhere they like, but in North America, many places are forbidden.) If you have the dream of camping on a beach, for example, find out which local beaches allow overnight RV visitors. Figure out which beach is the safest, most practical, and closest to a major city center.

You'll be an instant target for crime if you show up looking like you don't belong. Investigate local crime rates and research a city or town long before you actually roll up into it. Sometimes the most isolated parts of town are the most dangerous. (Tip: most poverty, homelessness, and crime festers in the easternmost parts of cities.) If you give off the appearance of a tourist who is brand new to RVing, you will attract the attention of people who will take advantage of you. Don't make yourself vulnerable. Educate yourself!

If you're not feeling brave enough to wander amongst the bears and cougars, you can get the best of both worlds by choosing a location near a city or town that simply appears more isolated than it actually is. Sometimes, in rural areas of North America and Europe, you can get the "falling asleep under the stars" experience without actually treading too far from actual civilization. Invest in a good-quality road map and figure out how many miles out you are willing to tread. And if you have a trusted, faithful canine companion, you'll definitely be safer.

Lastly, remember that the more prepared you are for a crisis, the more calmly you will be able to deal with it. If you're travelling with kids, tell them not to open the door to strangers or go wandering alone—especially at night. Keep your cell phone charged at all times, especially if you are leaving your RV while another person stays behind, or you are splitting up for any reason. Keep roadside assistance, such as AAA or CAA, on speed dial, and don't be afraid to call 911 if something seems amiss.

CHAPTER FOUR:
HOMESCHOOLING

If you have kids, many friends and relatives may recoil when you announce your plans to hit the road. Expect to face the standard question from well-meaning friends and family: "If you're going to be on the move constantly, how will they be able to stay in school?" The answer for many campers is to pull their kids out of school entirely, and allow the world to be their classroom.

History of Homeschooling

Homeschooling was once the standard model of education in the United States. The Puritans didn't doubt for a minute that it took "a village to raise a child," and children of all ages typically learned at home with their parents before they took on apprenticeships or headed to university. Education at that time was highly hands-on and devoted to practical application, since many children were involved their local communities at a young age.

Structured school days such as those common to today's

public schools, and compulsory school attendance laws, were not implemented until the early 19th century, when American educators learned of a popular German educational practice: the factory model school, known for its uniformity. Today, the vast majority of North American and European schools follow this standard.

Homeschooling as we think of it today did not become popular until the 1970s. At the time it was a fringe movement that was distinctly religious in nature. But over the past several decades, more and more families have chosen it as an alternative to traditional public schooling.

Today, there are more homeschoolers than ever before, and the percentage of homeschoolers is growing exponentially. In the United States, more than 2 million children are homeschooled. They are an increasingly mainstream and widely diverse group, ranging from rural homesteading farm kids to families living in the Bronx.

Some use religious textbooks that emphasize faith in all areas knowledge. Others use all-in-one curricula that mimics public school curricula, and encourage kids to take standardized tests like the SAT. Still others are "unschooling" RV travellers who allow their children to explore the world and discover their passions on their own terms.

Whether you choose to use a formulaic "school in a box" curriculum or prefer to let your adventures guide the daily lesson plans, your kids will surely reap the benefits of their unique upbringing. Many children would love the opportunity to travel around the world with their mom and dad at an early age.

However you choose to educate your children on the road, you can be sure that you are not alone. Far from being the controversial fringe movement it was in decades past, homeschooling is now cool, normal, and practical.

Myths and Benefits

Homeschooling myths remain prevalent despite statistical evidence that debunk them. Many people view the homeschooled child as an awkward, unseemly stereotype: devoutly Christian, sheltered, naive, asocial. Some people believe that homeschooling does not provide kids with the necessary preparation for higher education, despite the fact that homeschoolers continuously rank higher on standardized tests and obtain higher grades in college and university than their public schooled peers.

They seem to be faring better in the real world, too. According to *Business Insider*, homeschool graduates are generally happier and more involved in their communities than their peers. Many are accepted into high-ranking universities, including Ivy League institutions, which are becoming increasingly friendly towards homeschooling families.

Many universities feel that homeschooled kids have a wider range of knowledge than kids educated by rote in the factory model system, which helps them utilize their critical thinking skills better in rigorous undergraduate and graduate programs.

The prevailing belief that all homeschooling parents do so to protect their kids against "secular humanist" instruction and to ensure they remain strong in their faith as they grow up is not necessarily true, either.

While many parents do choose to homeschool their children for religious reasons, and use faith-based textbooks, there are just as many who choose to teach their kids at home for purely pragmatic reasons. They may prefer to instruct their kids in a less rigid and formulaic way. Most homeschooling parents state dissatisfaction with local schools, high rates of bullying and "cliquey" behavior, and a desire for more personal connection with their children as their reasons for pulling their kids out.

Cyberbullying has become a rampant problem for teenagers that teachers are unlikely to be able to offer a quick fix for. In some cases parents have found that school officials are unable to effectively combat bullying and have chosen to rehabilitate their kids in a safer environment away from it.

For these parents, homeschooling has helped their kids get back on track, mentally and academically, after experiences in the public school system that negatively impacted their mental health. Children facing extreme bullying, social isolation, learning disabilities, and mental illness often feel relieved to have an alternative offered to them.

There are academic benefits to homeschooling, too. Many homeschoolers graduate early. Virtually all of them have extra time in their days to devote to personal hobbies. It is not unheard of for high school homeschool students to take college-level classes in their spare time thanks to the extra hours in their day they can allot to their future goals.

They tend to be quite driven, mature, and eloquent in their speech. Contrary to the belief that homeschoolers are not properly educated, researchers have found that many enter college more prepared for the academic rigor expected by most universities.

Roadschooling

It makes sense that homeschooling has become the gold

standard educational model for RVers with kids. Where else can kids develop such valuable life skills, apply their practical knowledge, and socialize with people from all different backgrounds? There's never a dull moment for a youngster travelling across the country and meeting new people every day. Nomadic families are also more likely to raise bilingual children.

Kids educated on the road learn equally from textbooks and from the world around them. Homeschooling is ideal for nomadic families because kids can complete their daily allotted school time a lot more quickly than they would be able to in a classroom, and spend the rest of their day adventuring. In a sense, there is no better way to combine academic knowledge and application of real-world skills when your kids are seeing the world from the back window of a moving classroom!

Parents who travel frequently prefer having the option of keeping their kids close to them. There are many positive benefits to travelling with the little ones in tow. Parental involvement, especially during the first few years of life, is very important and crucial to childhood development. Close, tight-knit familial bonds prevent many social and psychological problems later on life.

You're unlikely to be the only camper in the park with kids. Frequent travellers will soon find that many other parents use homeschooling as an alternative to traditional schooling while on the road. Blogs such as Jennifer Miller's Edventure Project encourage and motivate young homeschooling, RVing families. In fact, it's become such a common phenomenon amongst campers that it has its own name: "roadschooling."

Roadschooling offers more flexibility than any other form of education for long-term travelling families. The percentage of formal versus informal schooling will vary from family to family, and ultimately, it will depend on your individual circumstances how you choose to implement a roadschooling curriculum into your family life. Some families emphasize hands-on "world schooling" and their kids spend less time hitting the books, while others prefer to stick to a more rigid schedule.

If your kids are very young, they might learn better with a looser, less rigid approach to education. On the other hand, if you're raising teenagers who are approaching college age, a long-distance education program might better suit their needs. This way, they can balance informal, adventurous learning with periods of intensive study that will prepare them for university or college.

As children age, they become more self-reliant and capable of taking charge of their educational goals. Most teenagers who were homeschooled for all or the majority of their lives will have a clearer idea in mind about what they want to pursue. If you are already a homeschooling family, making the transition to roadschooling will likely be simple, especially if you use a pre-organized curriculum.

CHAPTER FIVE:
STAYING CONNECTED

Adventure-hungry wanderers have always existed, but in the past, they have often bee marginalized or viewed as outsiders for their unconventional choice not to settle down at a permanent address. Stereotypes associated with nomads have historically ranged from mocking to downright demeaning.

The Desert Fathers of the early Christian church were social outcasts who chose a celibate, minimalist life steeped in prayer and contemplation in the great unknown. The hippies of the 1960s who lived out of their vans, preaching love and peace, were a laughingstock of their parents' generation. The first modern American homeschoolers in the 1980s created widespread controversy in certain states.

Millennials Have it Made

The good news is that in the current decade, people are generally less inclined to question you. Though there will always be a few naysayers who judge your choice not to participate in the rat race, there are an increasing number of people who are sympathetic, if not wholly supportive, of the decision to be a full-time RV traveler. And the rest? Well, they might secretly long for what you've chosen, too, but they will resist out of fear.

It isn't a secret to anyone that millennials haven't exactly inherited the best financial situation. *Time Magazine* reports that while most young people dream of owning homes, they find the possibility deeply unlikely. A combination of extreme disillusionment with the economy and a sense of hopelessness—a feeling that no amount of hard work will ever amount to anything—has driven many young individuals and families to reconsider their prioritizes.

For many, this has meant resigning themselves to being lifelong renters. It is admirable that millennials are putting in long hours even with little promise of upward mobility. But for others (including yourself, if you've read this far), the downfall of the American Dream is just too disappointing of a reality to face. Long-term RVers are giving up their ideals of home ownership in favour of something that sounds more financially doable and provides more opportunity for excitement and fulfillment: a life on the road.

There was a time when selling your house, dropping everything, and trading the commuter line for the rugged terrain meant certain things in regards to your social life. It was expected that you would lose touch with friends and family, due to the distance as well as the uncertainty about the next time you'd roll into a gas station with a pay phone. Up until the 1990s, you were lucky if you could send regular postcards to your family back home. But the Internet has changed everything so much that the modern nomad's life is nothing like the life of those who came before us.

They paved the way and made the thrill and unpredictable nature of the road life seem cool and appealing. But the simple fact of the matter is that we are living in a different age than they did. Today's RVers often make their choice out of frustration with the fast-paced digital time period we are living in. But in a strangely comforting paradox, many are also filled with gratitude for the fact that they are lucky enough to remain connected to their loved ones.

DIY Photojournalism

Take a quick glance at Instagram and you will see more than a few highly popular accounts run by individuals and couples alike. The photos are bright, slick, professional. The fashion is impeccable. The RVs are in beautiful condition. But the landscape? That hasn't changed a bit. It's still majestic. It still takes your breath away to witness it even through a photograph. The lay of the land hasn't changed, only the way we perceive it and interact with it.

Whether you're swilling scotch in a dive bar in rural Nevada, camping in the Everglades, or gambling in Dawson City, you can share photos and commentary on your experience with the press of a button. Not only that, but if you use social media effectively, you can monetize your adventures by blogging about them *and* inspire complete strangers to make similar life choices.

If you're a good writer, photographer, or both, you can create your own photo essays and share them with millions of people. There was a time when you needed to be in New York, London, Paris, or LA to reach so many people at once, but now, all you need is a relatively new smartphone and a wifi hotspot.

It isn't hard to find those, either. Even in rural areas you can remain connected to the Internet and many campgrounds provide access for a small fee. On the road you can stay updated on what your family is doing by keeping your Facebook account up and running. If you have an iPhone, you can easily FaceTime with family members. If you're concerned that your kids will miss out on precious time with Grandma and Grandpa, they can have face-to-face interaction from thousands of miles away. That is the age we are living in, and it's an amazing thing.

Of course, no matter how meaningful it may be, communicating online is still no match for real, human connections. It is important to keep your kids—and yourself—involved in social communities and meet new people no matter what town or city you visit. Engage with people from different backgrounds and walks of life. Get to know your campground "neighbours," even if you're only setting up shop for a few nights. Sit with them beneath the stars, share a meal with them, and learn their stories. Then, when you leave, add them on Facebook or Instagram and promise to keep in touch.

One of the most fascinating things about the social media age is how it allows people from all over the world to stay connected even after only one meeting. How many of us have heard our parents or grandparents wistfully recall a chance encounter with a friendly couple on vacation who they never saw again because they didn't get their address or phone number? This isn't a dilemma that millennials can relate to. If anything, social media has made it almost *too* easy to track down that nice person you met at the campground after you leave.

Some RVers truly wish to be liberated and free of all attachments to modern society. The minimalist movement is growing at exponential rates for this very reason. But one of the most common reasons that even the most savvy RVers stay on social media? They have friends in every corner of the world.

Sure, there is an element of narcissism involved in showing pictures of your exciting life on the road. But for most people who choose to stay "plugged in" despite living out of a small camper van in the woods, the motivation is clear: they want to know they have friends they can reach out to anywhere.

Once you have even one foreign friend, six degrees of separation starts to work its magic and suddenly you are linked in to many more people than you could ever imagine. If travelling abroad someday is one of your ultimate goals, this is an incredible benefit to staying connected to social media. Even if you only keep one account up and running, it reminds people that you're still there within reach, and gives them a quick and easy way to reach out to you if they happen to be in your neck of the woods.

Most people have a "dream vacation" in mind. Even if you've never spent a long time contemplating the possibility of where you would go because it seems unrealistic, there's a good chance that you could drum up the name of one particular place that sounds especially enticing to you. Maybe you've always longed to see the Coral Reef, build a school in Uganda, sip merlot in the French countryside, or ski in Colorado. Maybe the Norwegian culture fascinates you, or you'd love to visit a pub in the Scottish Highlands.

Whatever the case may be, if you open yourself to unconventional possibilities, you might find yourself a whole lot closer to achieving your seemingly impossible dream. At campgrounds all across the world, nomadic, ambitious, and fearless travellers just like yourself are dreaming similar dreams.

There might be seven billion people in the world, but your world becomes a lot smaller the more you see of it. So you could start out at a campground a few miles outside of your hometown and meet someone who knows someone who knows someone who has a housesitting opportunity in the Swiss Alps or the Upper West Side.

Without social media, you might have to go through a longer chain of connection before you ever reach those people and let them know you're interested in their offer. With social media, all you have to do is send them a quick email or register on Airbnb to prove you're a real person. Making connections worldwide is really that easy, and the more people you meet, the more people will be able to vouch for your trustworthy nature. Before you know it, you could be taking your travels a lot further than you could ever have imagined.

In 2017 and beyond, the call of the wild does not mean sacrificing your connection to the real world. It offers the best of both worlds. We have so much power in our handheld devices and the ability to use it to influence and inspire others. Why not use it to the best of our ability?

CONCLUSION

Everyone has different reasons for choosing this unique and exciting way of life over the standard "work all day, come home at night, get two weeks of vacation time a year" lifestyle.

It may simply be a matter of money and practicality. You may be a recent grad choosing to save up money for a few years and then settle down to buy your first home. You may be dissatisfied with years of office work or back-breaking labour, and you're ready to throw in the towel, call it quits, and say, "Enough is enough."

Additionally, your approach to travel will be unique because you are a unique individual. Everyone's experience is radically different and you don't have to feel compelled to live the glamorous, Photoshopped lifestyle you see on many popular Instagram travel couple's blogs. Your experience is yours and yours alone. You do not need to prove anything to anyone.

www.ingramcontent.com/pod-product-compliance
Lightning Source LLC
Chambersburg PA
CBHW021045180526
45163CB00005B/2296